THE Story OF You

Anna Claybourne

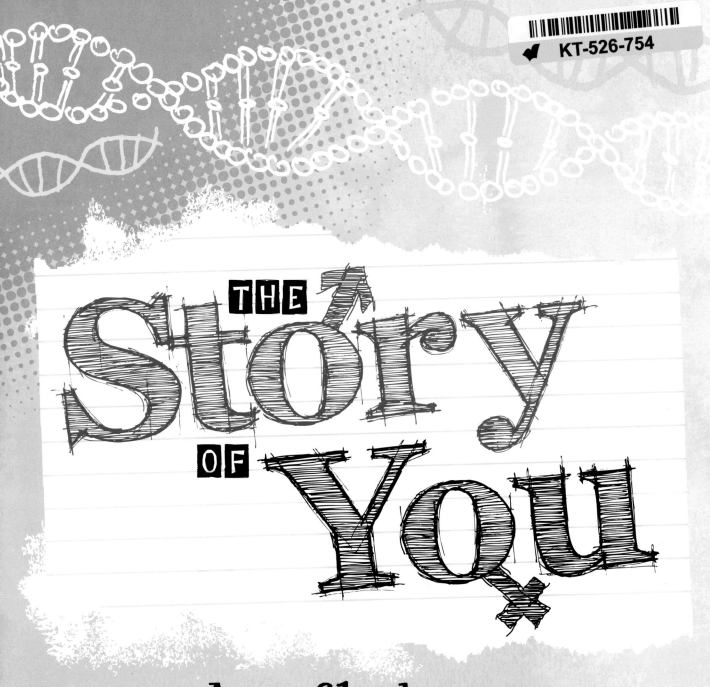

CONTENTS

YOUR PERSONALITY

YOUR FAMILY TREE

Who are you?

Wherever you go, whatever you do, one thing always stays the same — you're you! As long as you live, you'll be inside your body, looking out at the world, with your own thoughts and ideas, likes and dislikes, feelings, hopes and dreams. It's the same for all of us! But what exactly does "you" mean? What makes you who you are?

PARTS OF A PERSON

Each individual person is made up of various aspects and features. They're all different, but they're all part of you, and closely entwined together.

YOUR BODY

Everyone has a physical, breathing, moving body, and that's what keeps us all alive. Bodies come in many shapes and sizes, but we are all human beings and share the same basic body parts, like a heart, a brain, skin and blood.

YOUR PERSONALITY

Are you shy, or the life and soul of the party? Do you love swimming, strawberries or sci-fi — or hate them with a passion? Are you good with animals, neat and tidy ... or not? These are all parts of your personality — how you behave and what you're like.

WHAT'S IN A NAME?

When you're born, one of the first things you get given is your name. You don't get much choice about it, as you're a baby! Your name is yours to keep for your whole life, but some people do decide to change theirs. How do you feel about yours?

YOUR MIND

Your mind is the part of you that thinks and understands. Thinking is done by your brain, which is inside your head. Besides making decisions and having bright ideas, your brain stores your memories and all the stuff you've learned.

YOUR RELATIONSHIPS

From birth, we all learn about who we are, and our place in the world, through interacting with other people. Being "you" also means being someone's friend, cousin, son or daughter, pupil, team-mate or grandchild.

Nature or nurture?

You may have heard people talking about "nature and nurture". It's a quick, easy way of describing the two main ways you become who you are. So what does it mean?

NATURE

"Nature" means the things you have naturally, like your eye colour. Things like this are encoded in your genes and DNA – tiny parts inside the cells that make up your body. Your genes and DNA are passed on to you from your parents. They can also affect your brain, mind and personality. For example, being good at maths can be in your "nature" too. However, the genes you have don't decide everything about you on their own. That's where nurture comes in.

NURTURE

"Nurture" means the things that happen to you and change you during your life. The way you are looked after as a baby, the things you learn, and the foods you eat are all part of nurture — and they can have a big impact on who you are. If you're terrified of dogs, that could be because a dog scared you when you were tiny. If you grew up in a big, friendly family, that could explain why you're chatty and sociable.

WORKING TOGETHER

Nature and nurture are closely linked, and often work together. Take Jem, who's a very musical person. He's a great singer, can pick up tunes easily, and plays several instruments. Is that because of nature or nurture?

Musical skills like recognising notes and holding a rhythm can be carried in your genes. Most of Jem's family are musical, so he has music in his DNA — it's in his nature!

On the other hand, if he's from a musical family, that means Jem heard lots of music from an early age, and started lessons as soon as possible. Maybe he's musical because of nurture — the environment he's grown up in?

Of course, the answer is probably both — and it can be hard to tell exactly how much of each.

Your body

Where did you come from? Like everyone else, you started out as a single, tiny cell, created by your parents. That cell contained all the instructions it needed to grow into a whole human body. And not just any body, but your own, individual, unique body.

WHAT ARE CELLS?

Cells are the tiny units that living things are built from. Most of them are too small to see. But if you look at your arm, your tongue or your eyeball, you're looking at them! Millions and millions and millions of microscopic cells, all packed together.

Some simple living things, like these bacteria, have only one cell each. Larger living things have a lot more. A typical human body is made up of over 30 trillion (30,000,000,000,000) cells!

If you could look closely enough, you'd see that your skin is made up of cells.

MAKING COPIES

All types of living thing reproduce, or make copies of themselves — or, in other words, have babies. Your body was made when two cells, one from each of your parents, combined to make a new type of cell.

This special cell, called a zygote, divided in two to make two cells.

They divided again to make four cells, then eight cells and so on.

Eventually the zygote grew into a fully formed baby human — you.

BECOMING A BABY

EIGHT CELLS

DEVELOPING EMBRYO

BABY READY TO BE BORN

INSTRUCTIONS FOR LIFE

But how did that cell know how to grow into a human being? And how did it know how to grow into you — with your hair, skin, facial features and natural abilities? The answer is a chemical called DNA, which is found inside cells. It contains instructions for making cells work and grow the way they do. Each type of living thing has its own, unique DNA instructions.

When living things reproduce, they pass on a copy of their DNA in the cells used to make the baby. This explains why humans have baby humans, cats have kittens and an apple seed grows into an apple tree.

It also explains why children often look like their parents — because they get their DNA from them.

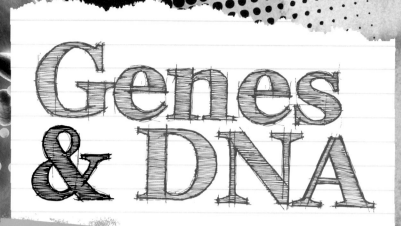

Genes & DNA

You'll often hear people saying something is "in your genes" or "in your DNA". But where exactly are genes and DNA — and how do they make you who you are?

INSIDE A CELL

cells, like the ones that [make up] your body, have several [...] [y]ou've probably seen [diagram]s like this before, [...] the inside of a cell.

short for [ri]bonucleic [acid — no] wonder we [...] "DNA"!

Organelles
(mini-organs that
do different jobs)

Cytoplasm
(jelly-like filling)

Nucleus
(the control
centre)

Cell membrane
(outer skin)

INSIDE THE NUCLEUS

But have you ever wondered how the nucleus controls the cell? The answer is your DNA, which is kept inside the nucleus. DNA is a chemical with a long, stringy, spiral shape.

Each DNA "string" is made up of four smaller chemicals called bases.

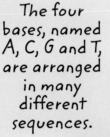

The four bases, named A, C, G and T, are arranged in many different sequences.

Each cell nucleus contains many spaghetti-like strands of DNA. All together, the DNA in one cell is about 2 m long.

READING THE CODE

A gene is a section of DNA, containing a sequence of bases that acts as a code. The code in each gene contains specific instructions for making proteins, the building blocks of the body. Teeth, hair, skin, blood — they are all built using proteins, so our genes are what make us work. When a body cell needs to do a job — such as making hair — it makes a copy of the instructions from the genes it needs. It uses this copy to put together the right ingredients in the right order.

As well as making body substances like hair, cells also build your body parts, like your face and your brain, as you grow. They follow genes to do this, too. So genes also decide things like the shape of your nose, and abilities like being musical.

LONG-DISTANCE DNA

If you stretched out all the DNA in all your trillions of cells, and joined it end-to-end, it would reach to the Sun and back — more than 10 times over!

Your human genome

Every type, or species, of living thing has its own genome. A genome is the complete set of genes and DNA needed to build a member of that species and make it work.

PASSING IT ON

When living things have a baby, that baby inherits a full set of genes from its parents. That means all the members of a species have the genome for that species.

Cats have a cat genome...

Bananas have a banana genome...

And humans have a human genome!

Cell copies its chromosomes.

Cell splits in two.

COPIES IN EVERY CELL

As you grow, your body makes new cells to build new body parts. New cells are made by older cells splitting into two. When a cell does this, it makes a copy of all the DNA in its genome, so that both the new cells can have their own copy.

This means that all your cells have their own copy of your genome inside them. They may not all need it. For example, the cells in your scalp that make hair don't need all the genes — they mainly just

CHROMOSOMES

The human genome is made up of 46 separate strands of DNA, known as chromosomes. There are 46 chromosomes in each cell nucleus.

In pictures, you'll often see chromosomes shown as lumpy X-shapes, like these:

In fact, chromosomes form this shape only when a cell is about to divide and they are being copied. The rest of the time, they are long, squiggly strands of DNA.

NO GENOME

There are actually some cells in your body that don't have a nucleus, and don't contain a copy of your DNA, such as red blood cells. They don't need it, as they do a simple job (carrying oxygen) and don't make anything. But most cells do contain a full genome.

HOW MANY?

Different living things have different numbers of chromosomes in their genomes. For example, a chicken has 78 chromosomes — many more than a human — but a mosquito has only 6.

In total, a human genome contains around 20,000 genes.

There's only one you!

If all humans share the same human genome, why are we all different? After all, there's no one else in the whole world who's exactly like you!

YOUR OWN GENOME

Though we all have the same basic genome, our genes can vary, and each person has their own unique set. For example, we all have genes for making hair. But there are genes for dark hair, red hair and fair hair, curly, wavy and straight hair. We all have different versions and combinations of these genes, so we all have different hair.

There are also some genes that not everyone has. For example, some people have a gene that helps them digest milk and dairy products. Other people don't have it, so they are "lactose intolerant" and eating dairy foods can make them feel ill.

MIXING AND MUTATING

One reason we are all different is that each person gets a mixture of genes from their two parents. You get 23 of your chromosomes from your mum, and 23 from your dad. Whenever two parents have a baby, they give them a different mix of their chromosomes. That's why even the brothers and sisters in a family aren't all exactly the same.

Another reason you are unique is that when cells copy their DNA to make new cells, they can make mistakes, causing the DNA to mutate, or change. Usually, these mutations don't cause any problems, but sometimes they can cause illnesses.

IDENTICAL TWINS

The only people who have the exact same genome as each other are identical twins (or other multiples) — because they started out as one cell, which then separated into two or more babies. But even they aren't totally the same, because of their DNA mutations. Nurture also helps to make twins different from each other as they grow up.

NO FINGERPRINTS

Most people have patterns of ridges on their fingertips — their fingerprints. But there are a few people in the world who don't have them. Their fingertips are smooth. Scientists have found that this can be caused by a single mistake in one gene.

Family traits

Many things about your body are decided by the genes you get from your parents, and can be different in different people. They are called "genetic traits".

TEST YOUR TRAITS

Here are a few genetic traits. Find out which you have, and if possible, see how many other people in your family have the same ones as you.

Photic sneezing

Does a sudden bright light make you sneeze? If so, you have a genetic trait called "photic sneezing".

What kind of earwax?

Is your earwax dry, flaky and golden, or wet, sticky and yellowish-brown? There are two possible types, decided by your genes.

Morton's toe

If you have Morton's toe, your second toe is longer than your big toe. Only about 1 in 10 people have it.

MENDEL'S PEAS

Scientists only discovered how genes and DNA work inside cells quite recently — in the 1950s and 1960s. But long before that, in the 1850s, a monk named Gregor Mendel did his own experiments, using pea plants in his monastery garden. He discovered how genetic traits get passed on and can skip generations. Mendel was really the first person to understand genes, although he called them "factors". It wasn't until much later that other scientists realised he had been right.

The traits Mendel studied included things like the the height of the pea plants, the shapes of their peas, and the colours of their flowers.

LOTS OF GENES

Some genetic traits are caused by just one gene being different. But most, like toe length and hair colour, involve several genes. The combination of genes you get from your parents decides what you end up with. Genetic traits can sometimes be passed down from grandparents to grandchildren, and "skip a generation" in the middle.

For example, Lyra has two grandmas, who both have blue eyes …

… while her parents both have brown eyes.

But Lyra, like her grannies, has blue eyes!

That's because Lyra's parents got a mixture of genes from their mums (Lyra's grannies), and their dads (Lyra's granddads). This combination of genes gave them both brown eyes. But by chance, they both passed on their "hidden" blue eye genes to Lyra, so she got blue eyes.

Nurture and your body

Genes play a big part in what your body looks like and what it can do. But nurture — the stuff that happens to you during your life — is very important too. It can change your body in all kinds of ways.

SUNLIGHT

Though too much sun can cause sunburn, we all need sunlight. It helps your body grow strong bones and fight off diseases.

SAFE SURROUNDINGS

Dirty, polluted air, a damp, mouldy home, or dangerous activities can all harm or even injure your body.

The forces of nurture

Whatever kind of body your genes give you, these things will all have an impact on it too:

NUTRITION

(getting enough healthy food) The food you eat affects your health, your weight and how tall you grow.

DAMAGING YOUR DNA

Some experiences can actually make harmful changes to the DNA in your body. Smoking — or breathing in a lot of smoke from fires — damages DNA in lung cells. Over time, this can cause illnesses such as lung cancer.

BIONIC BODIES

Sometimes a bad accident or illness can mean people actually lose a part of their body, such as a leg or hand. But it can often be replaced with an artificial or robotic one – a great example of nurture making changes to the body you were born with.

TRAINING

Can you ride a bike or type on a keyboard, and hardly have to think about it? Your body has learned these skills through lots of practice.

CHANGING YOUR LOOK

On top of the effects of nurture, many people make changes to their bodies to give them a different look. Some changes only last a short time, like dyeing, curling or straightening their hair. Others make a more permanent change, like having a tattoo, piercing their ears or having cosmetic surgery.

EXERCISE

Moving around a lot builds up your muscles and makes your heart healthier.

Your mind

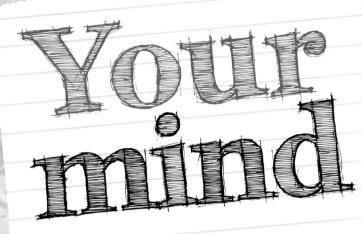

All the time you're awake, and even when you're asleep, your mind is busy. Thinking, planning, questioning and understanding, learning, imagining or dreaming — it never stops! So what, and where, is your mind?

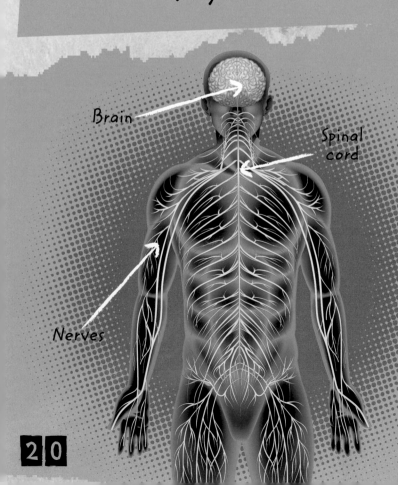

Brain

Spinal cord

Nerves

THE MIND AND THE BRAIN

We know that thinking, learning and remembering happen in your brain. And it certainly feels as if your mind is in your head. However, feelings like fear or excitement sometimes feel as if they are in your stomach or heart.

One way to think of your mind is as your "awareness" — all the things you are aware of feeling, knowing and remembering. You brain is connected to your spinal cord, and from there to nerves that reach all over your body. In this way, your whole body can "know" things, not just your brain.

ONE THING, OR TWO THINGS?

Some people say that the mind only exists because of what happens in your body. Ideas, feelings and dreams are just the result of brain and nerve cells doing their job.

For other people, the mind and the body are two separate things. You have a soul, or "spirit" that lives in your body, but isn't created by it — it's a separate thing.

There's no way of knowing which is right, though. What do you think?

CONSCIOUSNESS

In our minds, we know we are thinking, feeling and experiencing things — we are conscious. But how do cells in your brain conjure up conscious experiences, like seeing the colour orange, enjoying the taste of chocolate, or imagining yourself on top of a mountain?

The truth is, no one, not even scientists, knows exactly how the brain creates these everyday experiences — known as "qualia". We can't even be sure if other people's qualia — such as what chocolate tastes like or "orange" looks like — are the same as ours!

Your brain

Though some aspects of the mind are a mystery, we do know a lot about how the brain works. It has many parts that do different jobs, all helping to make you who you are.

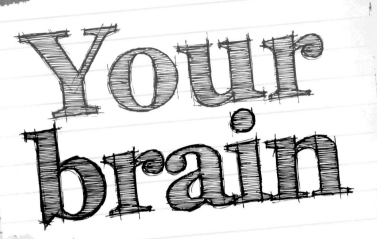

The cortex is the wrinkly grey outer layer of your brain. It's the part that thinks, understands and imagines.

The sensory cortex takes in touch signals from all over your body.

The frontal cortex plans things and makes decisions.

The corpus callosum links the two sides, or hemispheres, of the brain.

The hypothalamus helps to control your body temperature, hunger, thirst and sleep.

The hippocampus helps to store memories.

The brain stem controls basic things like breathing and heart rate. It also links your brain to the rest of your body.

PARTS OF THE BRAIN

In this picture you can see a brain cut in half, revealing some of its most important parts.

In case you were wondering, here's what your brain really looks like! All the thinking, dreaming, learning, remembering and imagining that makes up your mind is contained in a lump of pinkish-grey jelly, about the size of a small cabbage.

The thalamus sorts out signals coming into your brain.

CELLS AND CONNECTIONS

So how exactly does your mind think? The cortex is made up of billions of tiny, tree-like brain cells, or neurons. When you think, signals zoom around your brain, jumping from one neuron to the next, along many different, branching pathways. It's a bit like the way a computer chip sends signals around tiny electronic circuits. The brain is so complicated that there are trillions of possible pathways — which is why you can think, imagine and understand so many different things.

The visual cortex makes sense of the signals coming from your eyes.

MAKING NEW CONNECTIONS

What's more, when you learn new things, the brain actually makes new connections between its cells. So your brain has its own unique patterns and pathways, according to what you've experienced and learned. That's partly why your mind is unlike anyone else's.

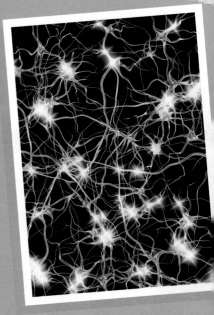

The cerebellum helps you move and balance your body.

23

Learning and intelligence

If you're a child or a teenager, you probably spend a lot of your time learning. But how does it work, and how does it shape your brain?

HOW WE LEARN

From the moment you were born, you started learning things like how to control your head, pick things up, eat, walk and talk. You learned words and facial expressions by copying other people. You learned to avoid things that were painful, like a hot candle flame. As you got older, you learned things like numbers, nursery rhymes or dinosaur names. For each thing you learned, your brain grew new connections, pathways and stored memories.

PLASTIC BRAIN

Children's brains are more "plastic" than adult brains, meaning they are better at changing and making new connections. This means that what you learn early in life plays a big part in making you who you are, as that's when you learn the most.

SLEEP ON IT

It's hard work for your brain, taking in lots of new information all the time. When you spend time learning, you also have to give your brain a rest by getting enough sleep. While you're asleep, things you have learned get stored in your long-term memory, so you don't forget them. This is one reason why babies sleep so much. It also means it's important to get a good night's sleep after a day at school.

DIFFERENT STYLES

Our genes can give us brains that work in slightly different ways. Everyone has some kinds of thinking and learning that they find easier, and others they find harder. For example, people with dyslexia tend to find reading and learning lists of facts quite difficult. But they are often very good at art and design, or problem-solving.

INTELLIGENCE

You probably know what you mean by "clever". But it's hard to say exactly what intelligence is, as it can involve many different abilities. They include:
- Learning how to do things
- Memorising facts
- Solving problems
- Coming up with creative ideas
- Understanding how things work
- Understanding other people
- Getting your message across clearly

How intelligent you are depends on both nature — the genes and DNA that gave you the brain you have — and nurture — your experiences and the things you've learned.

Imagination and ideas

You're up on the stage and everyone's cheering as you launch into your Oscars acceptance speech. Hang on — not really! You're actually in the middle of a maths lesson. What's going on?

MAKING IT UP

Our brains aren't just good at taking in and storing information from the world around us. They also have an amazing ability to make stuff up! Your mind has the power to dream up magical stories, original ideas, genius inventions and endless possibilities. It can imagine things that don't actually exist, and create characters, situations and whole other worlds. But what use is that?

DAYDREAMING

You might get told off for daydreaming sometimes, but it's actually good for you. Scientists have found that daydreaming helps your brain to work better. They aren't sure why, but it may help your brain to de-stress and focus more on separate tasks.

NIGHT DREAMING

Most people dream at night too, while they're asleep. No one is sure exactly why, but experts think it may happen as your brain sorts through your experiences, deciding what to remember and what to forget. Dreams often include real people and situations you may be worried about, such as moving house. They can give you a glimpse into what matters most to your mind.

One reason imagination is important is that it helps us have empathy — the ability to see things from someone else's point of view. If you're good at imagining, then you'll be good at thinking about how others feel. People with lots of empathy are often good at making friends, and at jobs like being a teacher or a vet.

Dreaming, imagining and thinking up ideas have been essential to human history. Without them, we'd have no inventions, art, songs, novels, TV shows or movies. Your hopes and dreams for the future are a huge part of who you are, and inspire you to aim high — even towards winning an Oscar, if that's what you want!

Your personality

What are you like? Your answer will probably be about your personality before anything else. It's a mixture of your attitudes and abilities, likes and dislikes, and how you deal with everyday situations — and it's an essential part of "you".

Your personality comes out all the time in what you do and say — but many people also like to express it in what they wear, their hairstyle, or the art or music they like — or even art or music they create themselves. Expressing your personality is important — it helps you to feel happy about who you are.

THE PERSONALITY CHARTS

Experts have identified lots of different aspects of personality, called personality "traits" (though these are not the same as the genetic traits on page 16). Here are just a few of them. Which of these describe your personality?

Nervous, a worrier	Brave, a risk-taker
Warm, chatty, loud	Quiet, shy or reserved
Emotional, easily upset	Calm and unflappable
Sensible, reliable, serious	Silly, giggly, funny
Relaxed, easy-going	Perfectionist, always in a hurry
Messy, forgetful, late	Organised and well-prepared

INTROVERT OR EXTROVERT?

One of the most important aspects of personality is the introvert–extrovert scale. Most people feel they are one or the other, though some are a mixture. Which do you think describes you best?

INTROVERT

- Being with other people tires me out
- I prefer to see one or two friends at a time
- I like reading and making things
- I work best on my own
- I love quiet, empty places

EXTROVERT

- Being alone bores and frustrates me
- I love big groups and parties
- I like chatting and team games
- I work best in a group
- I love noisy, busy places

BIG DIFFERENCES

If you think about your friends and family, you'll realise just how different people's personalities can be. While one person will happily try abseiling, another will be too nervous to do it – but will be great at making people laugh. You might have one friend who's brilliant in a crisis and always keeps her promises – and another who's always late, but doesn't bat an eyelid at singing or dancing on stage.

What do you love?

The things you love, and the things you can't stand, are a huge part of who you are. They affect how you spend your time, which friends you meet and even what you might do for a living...

Do you love Red?

Or prefer yellow?

Do you love to eat pasta?

Or prefer fruit and vegetables?

Do you love listening to pop music?

Or is rock, or indie, more your style?

Do you love to be outdoors?

Or snuggled up at home?

Do you love cooking and baking?

Or prefer to be outside, playing sport?

Do you love having a pet?

Or prefer to avoid animals?

A NEW HEART

Some heart transplant patients have claimed that after the operation, they suddenly liked the same foods or music as the person whose heart they had received – even if they never liked them before! Scientists are still investigating this, but it could mean that likes and dislikes are not just stored in the brain, but in other body cells too.

The human "yuck!" expression is the same wherever you are in the world.

THAT'S DISGUSTING!

Some smells, like rotten eggs and poo, are revolting to everyone. Humans have evolved to find them disgusting, because this helps us avoid dangerous germs.

But we also learn what is "disgusting" from other people. In some countries, most people can't bear the idea of eating a spider. In others, spiders are a popular food, but cheese is seen as gross. Children learn what is "disgusting" in their own culture by seeing expressions of fear and disgust on other people's faces.

Spiders for sale as a snack in Cambodia.

WHAT CAUSES LIKES AND DISLIKES?

Like many things, tastes like these are caused by both nature and nurture.

Twins and siblings are more likely to have the same likes and dislikes than people who aren't related ... even if they grow up separately. This means that genes play a part in deciding some tastes. For example, you might have a natural ability to understand musical notes, helping you to enjoy and play music.

However, your experiences have a big impact on what you like, too:

• Being good at something, like drawing or playing hockey, gives you a good feeling ... so you want to do that thing more (and vice versa!).

• Early experiences, even if you can't remember them, are important too. Maybe you like orange because you were fed in a cosy orange room when you were a baby.

• People often prefer what they are used to. So if you grew up in a football-loving family, or your parents always cooked Italian food, you'll probably like those things too.

You and other people

How often are you with other people — like your family, friends or schoolmates, or your sports team or orchestra? For most of us, the answer is most of the time! Your interactions with other people are very important in giving you a sense of who you are.

From the moment you are born, the way other people treat you shapes how you feel about yourself. In fact, it also shapes your brain.

Scientists have found that being loved, cared for and hugged a lot as a tiny baby helps your brain to grow the way it should. Through interacting with other people, we all learn a lot about how to behave.

SOCIAL ANIMALS

The reason other people are so important to us is that humans are social animals. We like to live in groups, and we survive by helping each other. Even people who live alone are part of their town or city, workplace and the society they live in.

IN SOLITARY

"Solitary confinement" means keeping someone in prison on their own, with no contact with other people. Around the world, it's one of the most feared punishments of all – being isolated and deprived of human interaction can have a terrible psychological effect on people.

GOOD AND BAD

Unfortunately, other people aren't always nice to us. A bad experience, like being bullied, can affect your personality – for example, by making you anxious, and wary about making friends. On the plus side, if you talk to someone about how you feel, and they listen to you, it can help you feel better again.

YOU IN THE GROUP

People's personalities have an effect on how they behave in a group. Are you a natural leader, or do you like to blend into the background? Are you the one people turn to with their problems, or for advice? Or are you the joker amongst your friends?

You and your memories

STORING IT AWAY

Each memory is stored as a pathway between a sequence of neurons in your brain. When you remember something, signals zoom along the same pathway as when it first happened. The more you experience the same thing – such as what an orange smells like – the stronger that pathway grows, and the harder it is to forget.

LOSING YOUR MEMORY

What if you lost your ability to remember faces, friends, words, or what you did this morning? This can happen, for example when people get brain illnesses such as Alzheimer's disease, or suffer brain damage. It can be very upsetting, because it's as if the person they were before has disappeared.

Colours, shapes and patterns

Tunes and lyrics to songs

Smells, tastes, sounds and textures, and what they all mean

Thousands of words, and the rules we use to make them into speech

I CAN'T REMEMBER!

Your brain can store a vast amount of stuff, but it's not perfect. Memories aren't always reliable, and some people are better at memorising things than others. Sometimes you know you know a word, but just can't remember it! Then, a few minutes later, it pops back into your head. Or you think you remember something happening to you, but it turns out it was a story you read. Confusing!

Like many other things, having a good memory is partly nature and partly nurture. You can be born with a brain that's good (or less good) at storing information, but you can also improve your memory with practice.

The human species

Most of us have our own family, including our parents, brothers or sisters, grandparents or cousins. But we're also part of a much bigger family — the whole human species.

ARE WE ALL RELATED?

In a way, yes! You might think of only your close family as your blood relatives. But experts think all today's humans are descended from quite a small group of early people, so we are all distant relatives. Wherever you go in the world, there are more similarities than differences between humans. We share lots of things, such as facial expressions, social behaviour and the way we form groups and communities.

WISE HUMANS

A species is a particular type of living thing. Scientists give each species its own scientific name, written in Latin, such as *Equus ferus* — the horse — or *Apis mellifera* — the honeybee. All the humans alive today belong to one species, which has its own scientific name, *Homo sapiens*. This is Latin for "wise human" — as humans have large brains and high intelligence, compared to most other animals.

Apis mellifera

Homo sapiens

Equus ferus

LANGUAGE

As you've seen in this book, being able to interact with other people is a very important part of who you are. And a huge part of that is language. Our ability to communicate in words lets us share all kinds of information, and pass it down from one generation to the next.

A vervet monkey on lookout duty.

TALKING ANIMALS

Some animals can communicate with each other, but as far as we know, they can only say a few simple things. Vervet monkeys, for example, make different noises to warn each other about different dangers, such as an eagle, snake or leopard. But they can't form long sentences.

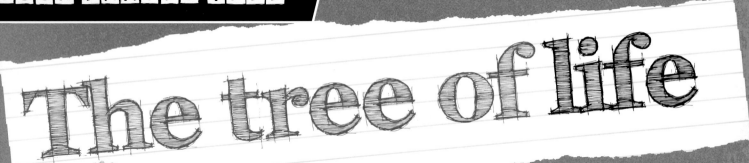

The tree of life

FOSSIL STORY

We can see how living things have changed over time from the fossils they left behind. For example, fossils reveal how horses used to be smaller and a different shape, but gradually changed into the modern species.

SHARED DNA

As you know, your human genome — the set of human genes and DNA — is what makes you human. But did you know that 98.8 per cent of it is the same as a chimpanzee's? Chimps are our closest non-human relatives, and our genomes are almost the same. Of course, you are less closely related to worms and bananas, but you still share a lot of genes with them. About 50 per cent of your DNA is the same as a banana plant's.

OTHER HUMANS

Imagine what it would be like if there were several species of human! There would be other animals that were humans, but not quite like us. Well, thousands of years ago, they did exist. Other human species included Neanderthals (*Homo neanderthalensis*) and Flores man (*Homo floresiensis*). But by about 12,000 years ago, they had died out.

Neanderthals had bigger skulls and brow bones than us.

Charles Darwin

Alfred Russel Wallace

The fossil skeleton of an early human ancestor, known as "Lucy", dating from over 3 million years ago.

DARWIN AND WALLACE

Two great scientists, Charles Darwin and Alfred Russel Wallace, figured out how evolution works at around the same time, in the 1850s. But it took many more years before we understood DNA mutations, and how they cause living things to change.

I think

This sketch from Darwin's notebooks reveals early ideas about the tree of life.

The human world

You're a human, thanks to the genes you got from your human parents. But what does that make you? How are humans different from other animals – earthworms, for example?

THINKING, INVENTING, BUILDING

Look around you! Humans have changed the planet more than any other species – which is why you're surrounded by buildings, roads, shops, farms and gadgets. Humans use their hands and their big, smart brains to create things. We've invented all kinds of stuff to make our lives easier, and we've worked out ways to build, manufacture or grow whatever we need. We construct huge cities

WORM
- Find food
- Have babies
- Avoid getting eaten by a blackbird

YOU

- Download that new song
- Text Naz about the party
- Write a book report
- Go shopping for trainers
- Football practice on Friday
- Remember Dad's birthday!
- Gran coming to visit at the weekend
- Concert rehearsal
- Need another eye test
  etc ... etc ...
  etc ...

Films Cartoons Design Art Games Jokes

Slang words **CULTURE** Talent shows

Stories Music Sports Recipes Vlogs

CREATING CULTURE

Humans are very good at passing on our knowledge and ideas. Whatever we invent, create or discover, we can share it with others, and write it down for people in the future to use as well. These shared, passed-on ideas are known as culture.

Over time, we have built up a complex cultural world – art, design, music, science, technology, politics, writings and beliefs that play a big part in our everyday lives. Our complex culture is partly why being a human is so interesting and exciting. And it's one reason why there's so much going on in your mind, personality and relationships.

THE STORY CONTINUES

As a human being, you are so much more than just a living body with a clever brain. You're packed with information, ideas, dreams, discoveries, love, friendship, experiences, passions, skills and ambitions. And you're only just beginning!

How will the story of you turn out?
It's up to you!

GLOSSARY

bacteria
Tiny, single-celled living things.

bases
Four chemicals found in DNA, which combine in different patterns to make genes.

cells
The tiny building blocks that make up living things.

cell membrane
The skin-like covering that surrounds a cell.

chromosomes
Strands of DNA found inside cells.

consciousness
The state of being awake and aware of yourself and your surroundings.

cortex
The outer layer of the cerebrum, the main part of the brain.

culture
The customs, ideas, creations and behaviours of a group or society.

cytoplasm

empathy
The ability to share or understand someone else's feelings.

evolution
The way species of living things gradually change over time.

extrovert
A sociable, outgoing person who loves company.

factors
A word once used to describe genes, used by the early gene scientist Gregor Mendel.

fossil
The remains or imprint of something that was once alive, preserved in rock.

gene
A section of DNA containing a pattern of bases that instructs a cell to do a particular job.

generation
The average length of time it takes for a species to reproduce and have offspring; or a group of people who are living at around the same time.

genome

lactose
A type of sugar naturally found in milk.

mind
The part of you that thinks, feels and has consciousness.

nature
Used to mean the natural features decided by your genes, such as eye colour.

nerves
Fibres in the body that carry signals between the brain and other body parts.

neurons
Cells that make up the brain, spinal cord and nerves.

nucleus
The control centre inside a cell, where the cell's DNA is stored.

nurture
Used to mean the way things that happen to you affect who you become.

organelles
Tiny, organ-like parts that do different jobs inside a cell.

organism
A scientific name for a living thing.

personality
The mix of traits, abilities, behaviour, likes and dislikes that are unique to each person.

physical
To do with the body rather than the mind; or to do with anything solid, 3D or "real".

plastic
In brain science, "plastic" means able to change and grow easily.

qualia
Experiences of what things are like, as understood by each individual.

reproduce
To have babies, or make more living things of the same species.

sociable
Friendly; good at chatting and getting on with people.

social
To do with society and groups, and the relationships between people.

species
A particular type of living thing.

spinal cord
A bundle of nerves leading from the brain down the middle of the back.

traits
Qualities or features that a person has, which can sometimes be decided by their genes.

zygote
A special cell, made by joining a male and a female cell together, which can grow into a baby.

FURTHER INFORMATION

What Makes You You?
by Gill Arbuthnott
Crabtree, 2016

*Evolution: The Whole Life
on Earth Story*
by Glenn Murphy
Macmillan, 2015

What is Evolution?
by Louise Spilsbury
Wayland, 2015

*The Usborne Introduction
to Genes & DNA*
by Anna Claybourne
Usborne, 2015

What Goes On In My Head?
by Robert Winston
Dorling Kindersley, 2014

That's Life
by Robert Winston
Dorling Kindersley, 2012

What Makes Me Me?
by Robert Winston
Dorling Kindersley,
2010

INDEX

First published in Great Britain in 2016 by Wayland

Copyright © Wayland, 2016

ISBN: 978 0 7502 9685 4
10 9 8 7 6 5 4 3 2 1

Printed in China

MIX
Paper from
responsible sources
FSC® C104740

Wayland
An imprint of
Hachette Children's Group
Part of Hodder & Stoughton
Carmelite House
50 Victoria Embankment
London EC4Y 0DZ

An Hachette UK Company
www.hachette.co.uk
www.hachettechildrens.co.uk

Editor: Elizabeth Brent
Designer: Grant Kempster

The website addresses (URLs) and QR codes included in this book were valid at the time of going to press. However, it is possible that contents or addresses may have changed since the publication of this book. No responsibility for any such changes can be accepted by either the author or the Publisher.

Picture acknowledgements:
Corbis/epa/Dan Levine 36; Getty Images/Arthur Tilley 4–5; Getty Images/Hero Images 6 (top left); Getty Images/LWA/Sharie Kennedy 7 (right); Getty Images 30; Getty Images/Nathan Dexter/AFP 31 (bottom left); Getty Images/Dorling Kindersley 39 (top); Getty Images/Mario Tama 41 (right); Science Photo Library 8 (bottom middle and right), 10 (top left), 16 (bottom), 23 (bottom right), 41 (top, middle left); Shutterstock.com/Joe Seer 26 (top); Shutterstock.com/Huang Zheng 28 (left); Shutterstock.com/Igor Bulgarin 29 (bottom); Shutterstock.com/Thomas La Mela 33 (top); Shutterstock.com/mikecphoto 34 (left); Shutterstock.com/Wayne0216 42 (left). All other images and background elements courtesy of Shutterstock.com.